LAB LOGISTICS AND SAFETY

BEAKERS, BURNERS, AND BALANCE SCALES!

EQUIPMENT IN THE LAB

ALISON AND STEPHEN ELDRIDGE

PowerKiDS press
New York

Published in 2021 by The Rosen Publishing Group, Inc.
29 East 21st Street, New York, NY 10010

First Edition

Editor: Jane Katirgis
Book Design: Reann Nye

Photo Credits: Cover Rabbitmindphoto/Shutterstpock.com; series art Triff/Shutterstock.com; p. 5 michaeljung/Shutterstock.com; p. 7 Pressmaster/Shutterstock.com; p. 9 (flask, beaker, graduated cylinder) urfin/Shutterstock.com; p. 9 (test tube) vipman/Shutterstock.com; p. 10 Africa Studio/Shutterstock.com; p. 11 PNOIARSA/Shutterstock.com; p. 13 (scale) Dudaeva/Shutterstock.com; p. 13 (pH paper) Anton Starikov/Shutterstock.com.com; p. 13 (microscope) netopaek/Shutterstock.com; p. 13 (thermometer) catshila/Shutterstock.com; p. 14 Pixel-Shot/Shutterstock.com; p. 15 nicolas_/E+/Getty Images; p. 16 urfinguss/iStock/Getty Images Plus/Getty Images; p. 17 angellodeco/Shutterstock.com; p. 18 mitsuap/Shutterstock.com; p. 19 monkeybusinessimages/iStock/Getty Images Plus/Getty Images; p. 21 Komsan Loonprom/Shutterstock.com; p. 22 wavebreakmedia/Shutterstock.com.

Library of Congress Cataloging-in-Publication Data

Names: Eldridge, Alison, author. | Eldridge, Stephen, author.
Title: Beakers, burners, and balance scales! : equipment in the lab / Alison and Stephen Eldridge.
Other titles: Equipment in the lab
Description: New York : PowerKids Press, [2021] | Series: Lab logistics and safety | Includes index.
Identifiers: LCCN 2019026878 | ISBN 9781725310261 (paperback) | ISBN 9781725310285 (library binding) | ISBN 9781725310278 (6 pack) | ISBN 9781725310292 (ebook)
Subjects: LCSH: Science rooms and equipment–Juvenile literature. | Science–Study and teaching–Juvenile literature. | Laboratories–Safety measures–Juvenile literature.
Classification: LCC LB3325.S35 E43 2020 | DDC 502.8/4–dc23
LC record available at https://lccn.loc.gov/2019026878

Manufactured in the United States of America

CPSIA Compliance Information: Batch #CSPK20. For Further Information contact Rosen Publishing, New York, New York at 1-800-237-9932.

CONTENTS

WORKING IN A LAB

Working in a laboratory can be lots of fun! In a lab, you can put all of your science questions to the test using the scientific method.

Every piece of **equipment** you'll find in a science lab has a certain job. To conduct a safe and successful experiment, you'll need to know what every piece of equipment does and how it can help you. You'll also need to know how to use the equipment properly so you can stay safe and avoid getting hurt. Let's take a look inside the lab!

IN THE LAB!

The scientific method is a set of steps scientists use to ask questions and make discoveries. The steps are: ask a question, form a **hypothesis**, test your hypothesis, draw conclusions, and tell others your results.

Different kinds of scientists may need different kinds of tools, but many tools are the same in every lab.

PROTECTION AND SAFETY

Safety is the biggest priority in any science lab. You'll need to put on your safety gear before you touch any of the equipment or **materials** in the lab because some of these things can be dangerous.

Gloves and aprons or lab coats will protect your hands and clothes from chemicals. Goggles make sure no chemicals or materials splash into your eyes. And you should always wear close-toed shoes.

A lab should have a first-aid kit in case someone gets hurt and an eyewash station in case something manages to slip past your goggles. It should also have a fire extinguisher, or a device that can put out fires.

Goggles, gloves, and lab coats protect scientists from the chemicals they're working with.

BEAKERS, FLASKS, AND TEST TUBES

A lot of the equipment in the lab is made of glass. Glass is great because you can see what's happening inside it. It's also good at resisting heat. But glass can break easily too. Before you use glass equipment, you should always check it for chips or cracks.

Labs usually have many types of **flasks**. A flask is a container that has a wide end and a narrow mouth. Labs also have **beakers** and test tubes. Beakers are **cylindrical** containers with flat bottoms that often have spouts, or places to pour, at the top. Test tubes are skinny tubes that are open on one end.

IN THE LAB!

Another container you might see in a lab is a petri dish. This special kind of dish is used to grow bacteria for experiments. Scientists have also used them to grow human parts such as ears and blood vessels!

Here are some different types of glassware used in a lab.

MIXING AND PREPARING

When you think of a science lab, you may think of scientists pouring different-colored liquids together. A lot of experiments *are* like this. But sometimes the materials you need for an experiment will need to be prepared before you can start.

A mortar and pestle is a tool that will help you grind things into a powder or mash them into a paste. Even if you're mixing liquids together, you may need to make sure they combine fully. A glass stirring rod works really well in laboratory glassware and is unlikely to have a bad **reaction** with your materials.

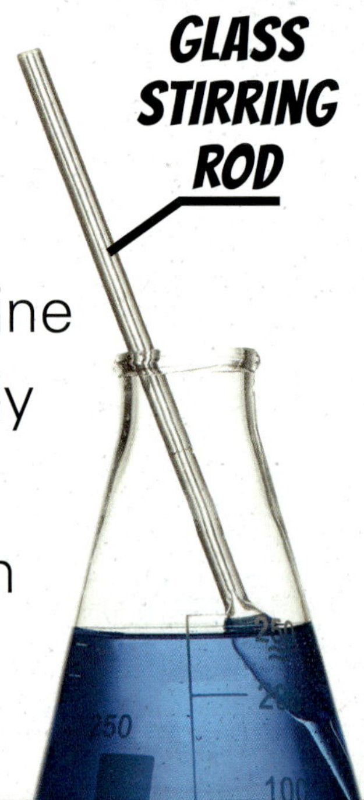

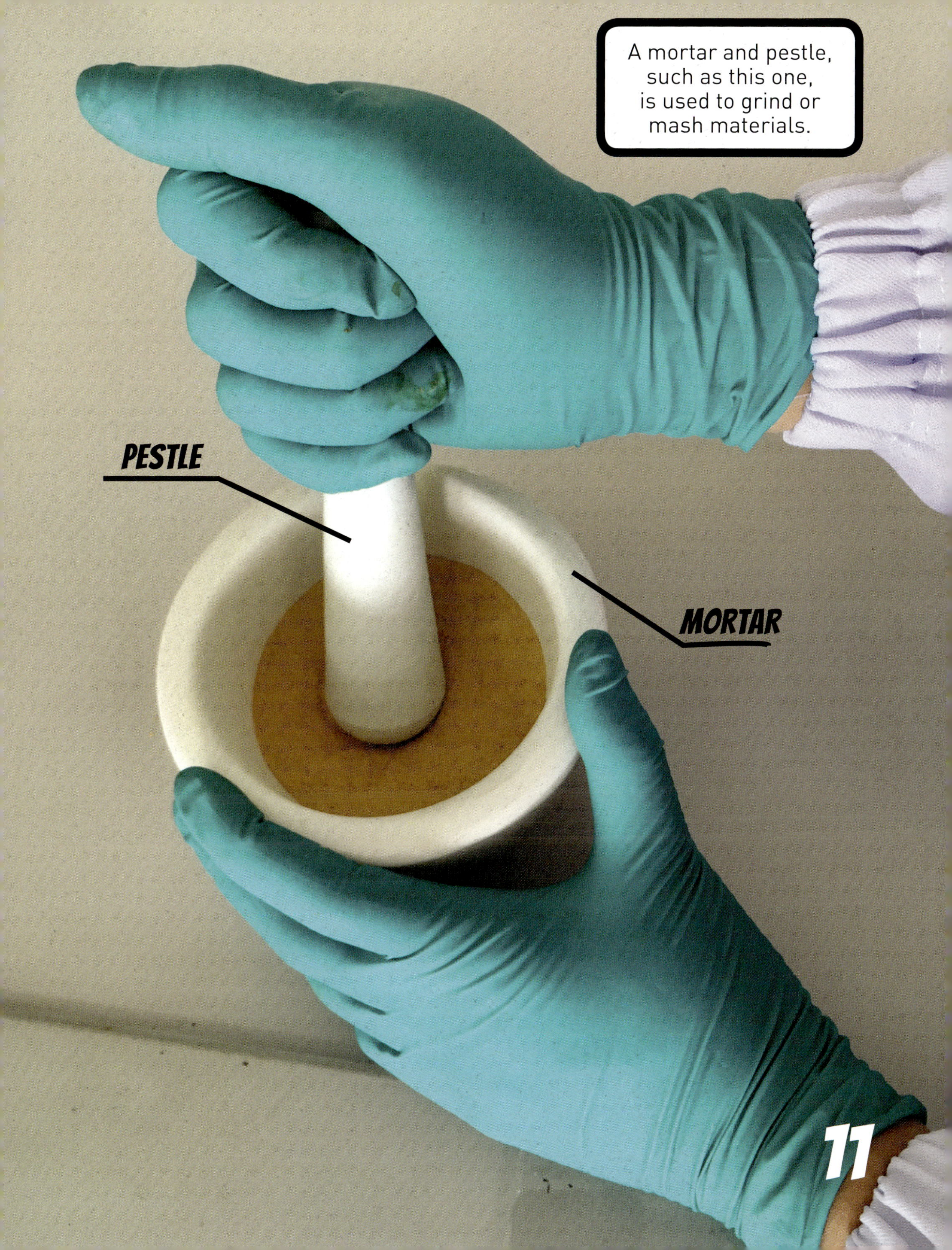

A mortar and pestle, such as this one, is used to grind or mash materials.

MEASURING

There are many reasons to measure materials in a lab. You may need a certain amount of a material for your experiment. You may want to see how a material changes during the experiment.

A graduated cylinder will let you measure how much of a liquid you have. A scale or a balance will let you weigh a material. A thermometer will tell you the temperature of something and whether that temperature increases or decreases. A microscope allows you to see and measure very small particles, and **pH** paper will tell you whether a material is an acid or a base.

IN THE LAB!

Specialized labs can be huge. The CERN laboratory houses the Large Hadron Collider, a 16.5-mile (26.6 km) ring that speeds up tiny particles and smashes them together so they can be studied.

MICROSCOPE

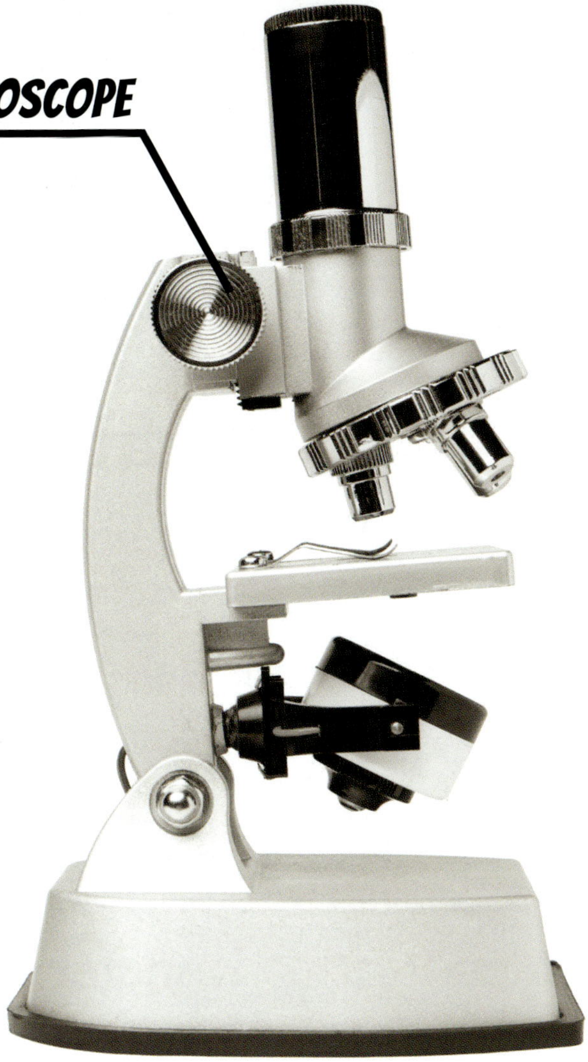

SCALE

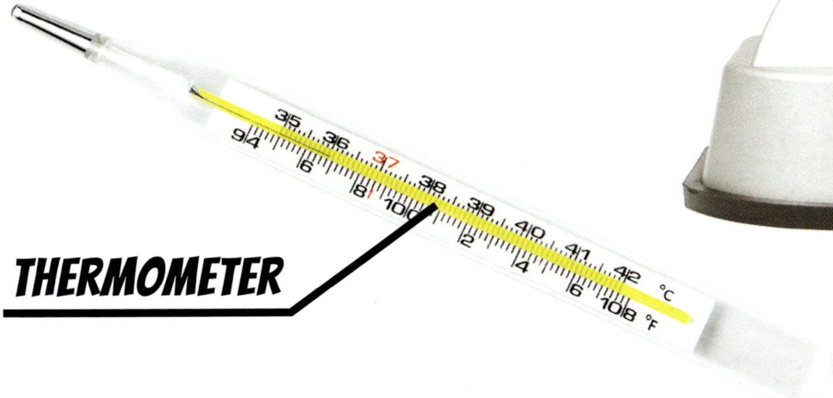

THERMOMETER

PH PAPER

PH paper changes color to show whether a material is an acid or a base.

TRANSFERRING MATERIALS

Often, while conducting an experiment, you'll want to move a material from one place to another. This should be done carefully. If you're using dangerous chemicals, you don't want any to spill or drip!

Funnels can help you pour liquid from one container to another. A pipette or dropper can be used to **transfer** a very small amount of a liquid. You can even use one to transfer a liquid one drop at a time.

A spatula can be used to move solid objects or materials. Make sure that, if you're moving something hot, you also wear heat protection for your hands.

Pipettes are also used to transfer medicine from one container to another.

HOLDING THINGS

Sometimes it can take a while for a reaction to happen. You may have to wait before you can find out the results of your experiment. So, it's good that a lab has plenty of equipment for holding things.

Labs that use test tubes will have racks that hold them. There are also many ways to hold tubes and flasks over flames for heating. You might use **clamps** or wire **gauze** on a ring stand. If you only need to heat something a little, you might hold it yourself using tools called tongs to keep your hands safe from the flame.

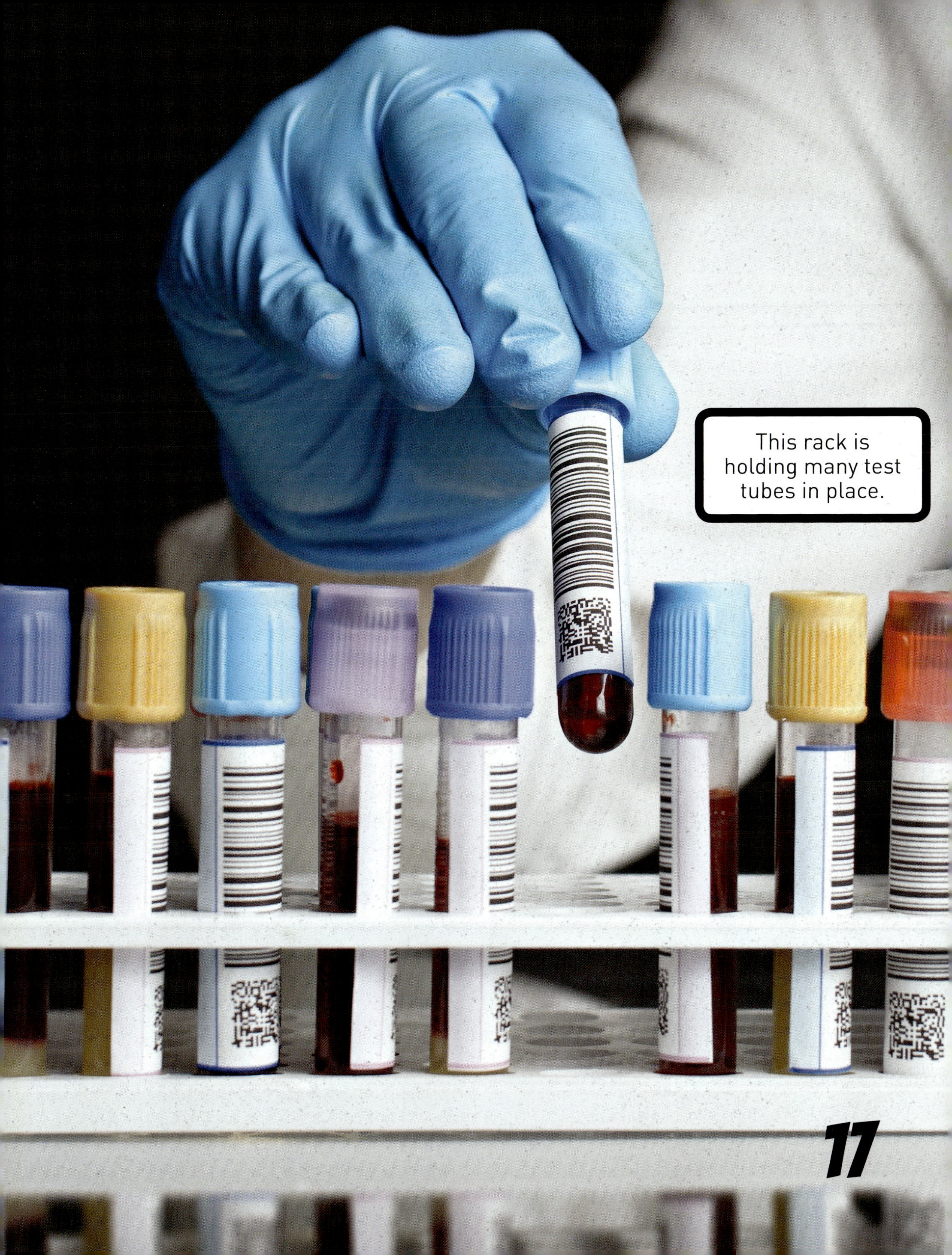

This rack is holding many test tubes in place.

HEATING

When you're doing your experiment, you might want to melt, boil, or even burn something. Or, you may just want to raise its temperature slightly. One way to do that is to apply even heat from an open flame.

One common type of device used to produce a flame in a lab is a Bunsen burner. A Bunsen burner makes a flame that is very hot and steady. To light a Bunsen burner, you can use a match or a device called a **striker**, which creates a spark that lights the burner's gas.

FIRE EXTINGUISHER

IN THE LAB!

When you are working with an open flame like the one produced by a Bunsen burner, always make sure you have a fire extinguisher nearby. Make sure your sleeves are away from the fire. If you have long hair, tie it back.

These student scientists are using a Bunsen burner to heat a liquid.

CLEANING UP

Cleaning up in the science lab is very important. Equipment that isn't clean could affect an experiment. It may cause a bad reaction. **Residue** from chemicals or powders left on tables or glassware might even hurt someone.

Luckily, soap and water are really good at cleaning most equipment in a lab. If you need to use a special cleaning product, your teacher will tell you. There are special brushes you can use to reach into test tubes and flasks.

You should always wear gloves to protect your skin when you clean your equipment. And when you're done, dry everything really well with paper towels.

IN THE LAB!

Sometimes you might need to clean up in the middle of an experiment or add water to something you're mixing. There's a special bottle called a water bottle for this purpose. It has a spout.

You can wash test tubes with a special skinny brush.

SAFE SCIENCE FUN!

Now you're ready to test your science hypotheses in the lab! What will you learn about first? Will you see what happens when you mix the liquid called vinegar with baking soda? Maybe you'll test the pH of lemon juice. Or maybe you'll look at pond water on a slide under a microscope.

With your new knowledge of science equipment, you can pick the right tools for the job. Put on your goggles, gloves, and apron and get to mixing, pouring, heating, and measuring. Just don't forget to clean up when you're done!

GLOSSARY

beaker: A wide glass with a flat bottom and a lip for pouring.

clamp: A small device that holds something tightly and firmly.

cylindrical: Shaped like a cylinder, or a solid object shaped like a tube.

equipment: Supplies or tools needed for a certain purpose.

flask: A glass bottle with a flat bottom and a narrow neck.

gauze: A very thin, loosely woven cloth.

hypothesis: A prediction about the answer to a question.

material: Something from which something else can be made.

pH: A number or scale for measuring whether a material is an acid or a base.

reaction: Something, such as a chemical change, that happens when two or more materials are combined.

residue: A small amount of something that remains after a process has been completed.

striker: A tool that creates a spark to light a gas flame.

transfer: To move something from one place to another.

INDEX

WEBSITES

Due to the changing nature of Internet links, PowerKids Press has developed an online list of websites related to the subject of this book. This site is updated regularly. Please use this link to access the list: www.powerkidslinks.com/lls/equipment